Art of Garnishing

Art of Garnishing

Inja Nam

Arno Schmidt

James Gerard Smith, *Photographer*

Van Nostrand Reinhold
New York

Library of Congress Catalog Number 93-7905
ISBN 0-442-01084-2

I(T)P Van Nostrand Reinhold is an International Thomson Publishing company.
ITP logo is a trademark under license.

Printed in Hong Kong

Van Nostrand Reinhold
115 Fifth Avenue
New York, NY 10003

International Thomson Publishing GmbH
Königswinterer Str. 518
5300 Bonn 3
Germany

International Thomson Publishing
Berkshire House, 168-173
High Holborn, London WC1V 7AA
England

International Thomson Publishing Asia
38 Kim Tian Rd., #0105
Kim Tian Plaza
Singapore 0316

Thomson Nelson Australia
102 Dodds Street
South Melbourne 3205
Victoria, Australia

International Thomson Publishing Japan
Kyowa Building, 3F
2-2-1 Hirakawacho
Chiyada-Ku, Tokyo 102
Japan

Nelson Canada
1120 Birchmount Road
Scarborough, Ontario
M1K 5G4, Canada

16 15 14 13 12 11 10 9 8 7 6 5 4 3 21

Library of Congress Cataloging-in-Publication Data
Nam, Inja, 1935-
        Art of Garnishing/ Inja Nam, Arno Schmidt; Gerard Smith, photographer
            p.   cm.
        Includes index.
        ISBN 0-442-01084-2
        1. Garnishes (Cookery) 2. Vegetables. I. Schmidt, Arno, 1937-
        II. Title
TX740.5. N35   1993   93-7905
641.8' 1--dc20   CIP

Contents

Individual Flowers

Flower Centerpieces

Foreword

Inja Nam is an artist whose work has earned our highest respect and admiration.

For nearly 20 years, she has practiced a unique culinary art form that has greatly contributed to our success at The Waldorf-Astoria. As Chef Hors d'Oeuvres in charge of creating all hot and cold canapés for banquet events at the hotel, she has inspired our entire staff with her imaginative and delicate work creating vegetable flowers and other innovative carvings and presentations. As Garde Manger, her decorative works have been on display at dinners of state, countless charity balls, and private parties that have entertained U.S. presidents and dignitaries at The Waldorf-Astoria.

With the publication of *Art of Garnishing* she shares this colorful tradition that has distinguished her years at The Waldorf-Astoria.

The staff of The Waldorf-Astoria joins me in congratulating Inja. We are extremely proud to have her on our team and look forward to her contributions to our hotel in the coming years.

Per Hellman
Vice President and Managing Director

Acknowledgments

When Inja Nam joined the Kitchen Brigade at The Waldorf-Astoria in 1974, I did not immediately recognize her special artistic talents. She is a tireless worker, skilled in aspects of garde manger (cold kitchen) work, easy to get along with, and always on time.

I cannot recall the particular occasion when we discovered the special talents of Inja. It must have been a very exclusive party given in The Waldorf Towers when Inja volunteered to "add a little decoration" to an hors d'oeuvres platter. This little extra decoration turned out to be a stunning bouquet of flowers carved from vegetables.

It did not take long before her vegetable flowers were in great demand. They have graced the tables of royalty, presidents, stars of stage and screen, prominent politicians and other world leaders.

Eventually, Inja Nam was dreaming about making a book of her art. After I left the position as Executive Chef of The Waldorf-Astoria I had more time on my hands, and I also learned how to work with publishers. It took Inja a number of years for her ideas to mature into a format with which I was able to work, and which I could present Pamela Chirls, Senior Editor at Van Nostrand Reinhold.

It quickly became clear that the success of the book depended heavily on the quality of the photographs. Pamela recommended James Gerard Smith, and thanks to his skill, understanding of the subject matter, patience, and stamina, all the pictures were completed in less than three weeks.

Taking the pictures at the hotel where Inja works made a lot of sense. We contacted Per Hellman, Vice President and General Manager of The Waldorf-Astoria and, after checking with the Executive Chef, John Doherty, Mr. Hellman graciously consented to allow all pictures to be taken at the hotel.

Most pictures of bouquets and displays were taken in banquet rooms, but we also selected some unusual locations. The Floralia flower shop in the hotel permitted us to take the Outdoor Party picture in the shop and provided the props. One picture was taken in the Silver Room of the hotel, where the silver is burnished and cleaned. The stewards were most accommodating and provided us with all the equipment we needed.

Once we needed a chandelier turned up, and Mr. Nash, the Chief Electrician, obliged immediately. There are many other employees who were always ready to help in the true spirit of hospitality, but special thanks go to John Doherty, my successor. He gave us the support we needed without asking questions. He suggested locations and magically got things done when we needed them done. Last but not least, I must thank Howard Karp, the Food and Beverage Director of the hotel, for his consent and encouragement.

Back home, I transferred my notes to the word processor. I am fortunate that my wife Margaret is a professional writer, gourmet cook, and gastronome. She found the time to edit my text and approached it with fresh eyes and an inquiring mind. Because I had spent 46 years working in kitchens, my original text assumed that all people would know what I meant. She took the approach that the book should be as reader-friendly as possible and assumed that many readers would welcome detailed instructions. Therefore, the text pages are very much her creation.

I would like to add my thanks to Mike Suh, Art Director at VNR, who designed this book, and to Chris Grisonich, the editor, who combed the text for inconsistencies and style.

Introduction

Vegetables and fruits provide some of the best materials for centerpieces, particularly for buffets or dinner-table settings. They convey a sense of plenty and bridge the gap between the inedible and edible.

Inja Nam, the author and artist, takes unpretentious vegetables and carves them into flowers to be used for decorative purposes. The flowers are beautiful, but basically they are very simple. The skillful combination of natural colors and shapes is the reason for their attractiveness. Vegetables have beautiful natural colors in many different hues, and when flowers are carved from vegetables, attractive bouquets can be created with little expense.

Flowers are fun to carve. The book provides four basic illustrated steps for each flower. In addition, the raw material for each flower, and the finished flower are shown in color photographs.

Materials Needed

All flowers were carved from vegetables available in most commercial markets, fruit stands, or supermarkets. Fresh fruits were occasionally used as components in centerpieces, but they were not carved. Many fruits tend to discolor (oxidize), and bleed when cut.

No artificial colors or materials have been used. Only toothpicks, bamboo skewers, and an occasional rubber band are utilized to hold the flowers and to provide support. In addition, some florist materials, such as rattan sticks, leaves, and wicker were used in the centerpiece bouquets.

The array of vegetables available in markets is increasing every year, partially due to food technologies. Better transportation, and the demands of our diversified culture also provide an impetus to market an ever wider selection of vegetables. Not long ago, for example, bell peppers came to the market green, with a limited supply of red ones available. Today, bell peppers are also grown in yellow, orange, and purple. Many supermarkets carry Oriental and South American vegetables and ingredients such as yucca, jicama, daikon radish, and Chinese chives.

We were wisely encouraged by the publisher to use only commonly available ingredients, to make this book most practical to a wide audience. Thus, nearly all vegetables are available throughout the U.S. in all seasons. Only a few of the ingredients may be harder to find in some parts of the country, but they should be available with a little searching. To fill out some bouquets pictured in this book, common vegetable market products, such as Chinese or regular size chives, garlic chives with blossoms, artichokes, acorn squash, and other items were used. In some bouquets, hidden in the vases and bowls, were large eggplants and squash to provide a stable base for the flowers.

Some bouquets include lemon and galax leaves for added dramatic effect. Both are easily obtainable from all florists and are nontoxic. When not available, other leaves can be used according to season. Stay away from lettuce leaves and parsley, however; both tend to wilt rapidly. Flexibility, good taste, and creativity are the guiding forces when creating vegetable display pieces.

All vegetables should be firm and fresh. It is important to buy enough to have a number of pieces of each variety to be able to choose and to reject pieces that are not suitable.

Vegetables are natural products and sometimes can harbor surprises when cut open. Leeks can be tough and woody, kohlrabi netted with leathery fibers, turnips mushy, and squash riddled with insect holes.

Vegetables come in many sizes and are often governed by season, price, and places where they are grown. The sizes of carved vegetable flowers in most cases are governed by the original sizes of the raw materials because the color of the vegetable skin is an important component. Some vegetables can be trimmed to make smaller flowers when the skin is not shown.

Tools

Very few tools are needed. Basically, a sharp paring knife, a French knife, melon scoops in various sizes and, of course, a sturdy cutting board are necessary.

Many flowers start with thin slices. They are best made on an electric meat slicer, available in virtually all commercial kitchens. Households rarely have an electric meat slicer, but some have an old-fashioned manual bread slicer. This will do, providing it can produce really thin slices. In some cases the slices can be made with a French knife, although even an experienced cook will have trouble hand-slicing certain vegetables thin enough to roll and wrap around a center as some instructions require.

Some kitchens have a tool called a mandolin. It is a manual vegetable slicer with adjustable blades. It can be used to make thin slices, but fingertips often get nicked in the process. Carrots can also be used in food processors, but not larger vegetables such as turnips. If a meat slicer is not available, make flowers that are not based on thin slices.

Selecting the Components

Inja Nam has taken advantage of the wide variety of vegetables available in supermarkets, farm stands, and ethnic food stores.

She is a skillful shopper when selecting vegetables; she can picture in her mind how a certain vegetable on the shelf will fit in her bouquet. She can also recognize the freshness of the produce and to what extent it might be tough, woody, or hollow. If you are not skilled in vegetable shopping, buy extra and use the leftovers in cooking.

Knife Skills

Inja has spent a lifetime in professional kitchens and is obviously very skilled working with knives. However, none of the flowers described are so difficult that they cannot be duplicated by someone not as skilled. It might take a little practice and perhaps a few Band Aids, but basically the flowers are easy to make. This is by design; we did not want to make a book that requires an accomplished artist and a master carver to make the flowers.

Basically, the vegetable flowers in this book are easy to make when the directions are followed. Imagination and a willingness to practice are all that is needed. Keep in mind that nature is not perfect and unevenness, varying sizes and color shades, and even imperfections add to the attractiveness of the displays. Trying to be too perfect does not look natural. The flowers should not look like machine-made plastic flowers.

Just get started. Do not be too fusy right away. Have fun with it. You will be surprised how well they turn out.

Finding the Right Props

Inja Nam has a lifelong collection of stunning vases and bowls, many of which she has used as props in this book. However, even without Inja's collection, beautiful displays can be created. In addition to regular vases, punch bowls, china bowls, baskets, even old cooking pots, suitable props can be found at flea markets and garage sales.

Harmony in proportion is important. The flowers must be in proportion to the container, the size of the table or niche where it is set and other aspects of the display. A certain bowl may not work for a bouquet you have created, but keep it and let it inspire you for the next one. The bouquets in this book depict seasons and occasions, but most displays can be used almost any time. It is wise to remember that certain vegetables are associated with seasons, and although available, they should not be used.

Displays

The vegetable flower displays are not meant to be eaten. They are assembled with sharp toothpicks, pointed bamboo skewers, and occasionally with rubber bands. It is very important to keep the displays out of reach of guests, and to inform the service staff that the flowers are not the vegetable crudité on the buffet. Terrible accidents can happen if a guest gets injured by a hidden toothpick while eating a flower. The edible crudité should be well separated from the other flower centerpieces.

Some pieces, especially when made with onions and garlic chives, can have a rather strong odor. They should not be used as table centerpieces when people are actually seated around them. They are better suited as buffet displays

We do not recommend mixing fresh flowers with vegetable flowers. The contrast will likely detract from the artistry of the vegetable versions. The vegetables are carved to look flower-like, yet should be recognized as what they are — this is part of their fascination.

Preserving the Display

Heat and low humidity are the enemies of vegetable displays. Vegetables dry out fast. Many individual flowers, or flower components, can be made ahead of time if they are kept in ice water in the refrigerator. The step-by-step instructions for individual flowers normally indicate how they can be stored. Once assembled, they can be sprayed with cold water. Any clean spray bottle will do nicely. If the display is completely out of reach of the customers, and there is no chance whatsoever that the display will be eventually eaten, the spray can be mixed with one part glycerin and four parts cold water. The glycerin is not in itself toxic, but it should not be eaten. Glycerin makes the droplets stick to the surface and the display will look fresh for a longer time.

The flowers can also be kept tightly wrapped in plastic film. This is useful for off-site catering, when the material must be transported without water. They should be sprayed frequently with cold water when unwrapped.

Waste

Vegetables are inexpensive when compared with live flowers, and the trimmings should be saved and used. Good cooks never throw anything away, and all leftovers can find uses. It is important to adopt clean working habits. You should work with one type of vegetable at the time, and carefully separate useable trimmings from waste during the work. This method makes cleaning up easier and saves much time.

Photography

All photographs in this book were taken at The Waldorf-Astoria Hotel in New York City. The present building was opened in 1931, is a historic landmark, and is probably the most stunning public edifice in the Art Deco style in New York. The hotel is beautifully maintained by the owners, Hilton Hotels Corporation, and it was not difficult for our team to find the proper location for each display.

The photographer for all pictures was James Gerard Smith. Jim has a contagious enthusiasm for his work, unending patience, and a quest for perfection. This led to long shooting sessions. Every morning, the bouquet shots were briefly discussed with Inja, and their size and theme were assessed. Then Jim and I went on a location hunt to find the right background to compliment the size, colors, and theme of the displays. The location hunt took us from the cellar to the attic of The Waldorf Towers. All pictures were taken in August when the hotel activities are lighter than in other seasons. When we found the right spot, we sometimes had to move a piece of furniture or request a special background prop.

As soon as the location was selected, Jim and his assistant had to transport the bulky equipment to the site and find electrical outlets or enlist the services of the chief electrician to turn on the lights. I had to transport the piece to the site. Since I spent a decade working at the hotel as Executive Chef, I knew the intricacies of the infrastructure, the different freight elevator locations, and landings and passageways. Still, each photo session took many hours and there were some anxious moments.One picture was taken in front of the Ballroom elevator doors. The Ballroom was closed, and we did not expect any interference when we photographed in the Silver Corridor in front of the Ballroom elevators. Unexpectedly, the doors opened and a startled tourist almost fell on the display and the camera equipment.

We took the Thanksgiving picture on the outside terrace off the eighteenth floor Starlight Roof. We picked an overcast day to simulate Thanksgiving weather in New York, but got more than we bargained for because it started to rain during the photo shoot.

The step-by-step photographs were taken in a section of the butcher shop in the main kitchen. The place was air conditioned, which made the long sessions bearable. To get the shoots at the right angle, Jim had to stand on a stepladder, and Inja had to slowly repeat each step just to make sure we got the right sequence. It was a lengthy process.

Individual Flowers

Acorn Squash Flower

Materials Needed

One small to medium-size acorn squash,
one carrot,
one blackberry, and
a bamboo skewer.

Tools Needed

An electric meat slicer, a paring knife, and garden clippers.

With the electric meat slicer set at #2, cut the squash into horizontal slices. The slices must be solid in the center. Since the squash has a cavity, only the first slices from the top and bottom can be used.

Place 4 to 5 slices on top of one another and arrange them into an attractive flower. Alternate the grooves and the pointed tips to create the petals.

Cut one horizontal slice from the wide end of the carrot approximately 1/4" thick. This will be the base. Insert the bamboo skewer through the carrot slice and through the slices of squash.

Arrange the petals on top and place the blackberry in the center. With the garden clippers, trim the skewer to a desired length.

Comments

When blackberries are not in season, most other small berries will do. Cranberries make an attractive alternative and are in season during the winter. This flower was photographed with lemon leaves.

Tricks

Acorn squash has a tough but attractive exterior in hues ranging from dark green to orange. By skillfully alternating the colored edges, a wonderful effect can be achieved.

Beet Rose

Materials Needed

One firm, medium-size beet, a bamboo skewer, and lemon leaf stems (optional).

Tools Needed

A vegetable peeler and a paring knife.

Peel the beet.

With the tip of the paring knife make five or more vertical cuts around the perimeter of the beet, but do not cut all the way down. These cuts will become the outside petals.

Carefully make parallel cuts inside the first cuts to form slender V-shaped cuts. Remove the layer of vegetable material inside the V-shaped cuts. Make sure the petals remain attached.

Make five or more V-shaped cuts inside the first layer of petals. These cuts will form the next layers of petals. Again remove the thin layer of material inside the petals. Continue this procedure until the center is reached. The petals should become progressively smaller as you reach the center. The center consists of a cluster of small petals.

Comments

A strong bamboo skewer is needed to support this heavy flower. Beet color is very strong. It is water soluble and will stain anything upon contact. It is important that the vegetables are firm and fresh.

Tricks

To help remove beet stains from your hands and equipment, use lemon juice. By using various sizes of beets, attractive bouquets can be created consisting only of beet roses. This same method can be used to create turnip flowers.

Beet Tulip

Materials Needed

One firm, medium-size beet, one pearl onion, a bamboo skewer, and several thin leeks or firm scallions.

Tools Needed

A paring knife and a melon scoop (optional).

Place the beet on a cutting board. Cut off the root end to create a stable base. Make five slightly slanted vertical cuts to create a five-sided cube.

Make five vertical cuts into the five edges of the first cuts. This will create pointed petals. Be sure not to cut all the way down; the petals must remain attached at the base.

$\mathcal{C}$arefully remove all of the center material.

$\mathcal{S}$hape and round off the petals and base as needed. Insert a bamboo skewer from the bottom up and place a pearl onion in the center.

$\mathcal{C}$omments
Beet color is very strong. It is water soluble and will stain anything upon contact.

$\mathcal{T}$ricks
To partially remove beet stains from your hands and equipment, use lemon juice. A large melon scoop can be used to remove the center core.

Carrot Paint Brush Flower

Materials Needed

One large-size carrot, as straight and firm as possible, sturdy tooth-picks, and *kanpyo* (thin Oriental dried gourd strips).

Tools Needed

A vegetable peeler, an electric meat slicer, a paring knife, and garden clippers.

Lightly peel the carrot (optional). With the electric meat slicer set at #2, slice the carrot lengthwise. Overlap ten slices on the cutting board and trim the edges evenly. The slices should be slightly narrower on one end. The bundle should be tapered to resemble the original shape of the carrot.

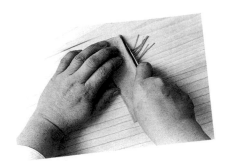

While keeping the bundle together, make cuts on one side, as close together as possible, with the paring knife. The cuts should extend to approximately one-third of the width of the carrot bundle.

8

$\mathcal{S}$tart at the narrow end of the carrot and roll the first slice into a tight roll. Roll the other slices around the first slice and continue this procedure until all of the slices have been used. Make sure that the slices are rolled tightly. Secure the pieces with toothpicks while you work.

$\mathcal{T}$ie the bundle tightly with *kanpyo* strips. Clip off the protruding toothpicks.

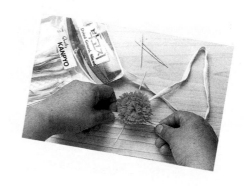

Comments

Kanpyo is a natural Oriental product made from dried gourds. It is available in Oriental markets. If not available, use rubber bands, which should be hidden by tying chives or thin leek leaves over them.

Tricks

The carrot flower will curl in ice water, but *kanpyo* will soften in water after approximately two hours. The flowers without *kanpyo* can be kept in ice water for two to three days.

9

Carrot and Leek Gladiolus

Materials Needed

One thick carrot, one very long, thin leek with many branches, and sturdy toothpicks.

Tools Needed

An electric meat slicer, a paring knife, and garden clippers.

With an electric meat slicer set at #2, cut the carrot into horizontal slices. Shape the slices into cones and secure them with toothpicks.

Trim the leek branches into attractive shapes. The objective is to provide sturdy, evenly shaped branches.

Arrange the blossoms evenly on the stem. With the garden clippers, clip off the protruding toothpicks.

Continue arranging the blossoms on the leek stems. Vary the sizes, using small blossoms on the top and large blossoms on the bottom.

Comments

Very large-size carrots are often tough and woody in the center, and they are difficult to roll. The shape and size of the leek stem is important for the successful creation of this flower. The long leek used in this picture was purchased at an Oriental vegetable market.

Tricks

The gladiolus flower can also be made with ginger blossoms. (See page 20.)
A combination of carrot and ginger blossom gladiolus make a nice bouquet.
It is best to work with the slices at room temperature. If the slices are kept in ice water, they will become stiff and difficult to roll.

11

Carrot Rose

Materials Needed

One firm, large-size carrot, sturdy toothpicks, *kanpyo* (thin Oriental dried gourd strips), and lemon leaf stems.

Tools Needed

An electric meat slicer, a paring knife, and garden clippers.

With an electric meat slicer set at #2, cut eight to ten thin round carrot slices. Roll the smallest slice into a tight roll to create the center.

Add petals around the center, pinching the base to form a shape like a rose petal. Secure the work with toothpicks as it progresses.

12

$\mathcal{T}$ie the flower with *kanpyo* and with the garden clippers, clip away the ends of the toothpicks. Leave only a small portion of the toothpicks still protruding, hidden by the petals, to keep them secure.

Comments

Kanpyo is a natural Oriental product made of dried gourds. It is available in Oriental markets. If not available, use rubber bands. The rubber bands can be hidden by tying chives or thin leek leaves over them. Lemon leaf stems look very natural because they resemble rose stems (without thorns!)

Tricks

The flower can be kept in ice water without *kanpyo*; *kanpyo* will soften in water after approximately two hours.

Carrot Tiger Lily or Tulip

Materials Needed

One firm, large-size carrot, green peas, sturdy toothpicks, and a wicker stick.

Tools Needed

A paring knife.

Point the root end of the carrot away from you. Start at the middle of the carrot and make four to six cuts approximately 3" to 4" long toward the tapered end. Cut off the carrot point approximately 3/4" below the cuts to form the bottom of the flower.

With the point of the knife, make a second set of cuts into the first ones inside the carrot to shape the petals. Be sure petals remain attached. Cut into the center, but do not cut all the way to the bottom. Leave about 3/4" of the material inside. Be careful not to break off the petals.

14

*T*wist out the center and discard.

*U*sing toothpicks, attach a number of peas to the center cavity. The number of peas depends upon the size of the flower and the effect desired. Place the flower on the pointed wicker stick.

Comments

This flower is best displayed when surrounded with large, fresh chives.

Tricks

Twisting out the center might be difficult without practice. To remove the material, you may want to use a small melon scoop. Carrots are hard, and spearing them with wicker sticks may be difficult. If necessary, make pilot holes with a sharp bamboo skewer.

Eggplant Large Flower

Materials Needed
One medium-size purple eggplant, one firm, small-size leek, and a bamboo skewer.

Tools Needed
A paring knife.

Place the eggplant on a cutting board with the stem end pointing away from you. Carve long, slender petals with the tip of the paring knife and be sure the knife reaches into the center.

Carefully separate the halves.

16

*S*crape and clean the inside and open the flower.

*C*reate a leek flower for the center. Cut off the roots as close to the base as possible without separating the leaves. Discard the roots. Cut off the top of the leek approximately 1 1/2" from the end of the root. Make several incisions from the top almost to the base. Spread the leaves apart. (For additional details, see Leek Paint Brush Flower on page 24.) Insert the skewer through the stem of the eggplant and secure the leek into the center.

Comments

*B*oth the top and bottom parts of the eggplant can be used, but the stem part is more elongated and therefore more elegant. The stem also provides a secure base for the bamboo skewer. The beauty of this flower is determined by matching the size and color of both the eggplant and the leek. The eggplant should be firm and a shiny purple. The leek should be small, firm, and slightly green.

Tricks

Eggplants oxidize rapidly. The eggplant part of the flower should be dipped frequently into lemon water as you work to prevent discoloration. The leek center should be placed in ice water to keep it open. The eggplant should not be kept overnight in ice water because it will get soggy.

Endive Flower

Materials Needed

One firm, medium-size Belgium endive, one white turnip, toothpicks, and a bamboo skewer.

Tools Needed

A paring knife and garden clippers.

Carefully peel away the outer leaves of endive without breaking them. Save the leaves. The center core should remain intact.

Cut the turnip into small cubes or triangles approximately 1/8" thick and 1/2" across. These will support the leaves when assembled.

Arrange leaves, turned inside out, around center core as follows: Spear each turnip triangle with a toothpick, add on an endive leaf, and fasten to the center core.

With the garden clippers, trim off the protruding toothpicks, and insert the bamboo skewer into the bottom.

Comments

Endives are a luxury salad vegetable imported from Belgium and Holland. The height of the season is winter. Do not soak this flower in ice water.

Tricks

This flower looks best in a large arrangement. Its color and size provides an attractive contrast with surrounding flowers.

Ginger Blossom Gladiolus

Materials Needed
Ginger blossoms
(*miyoga*), one firm,
small-size leek, and
sturdy toothpicks.

Trim the leek branches into attractive shapes. The objective is to provide sturdy, evenly shaped branches. Spear each blossom with a toothpick and fasten it to the top part of the leek stem. Trim the leek branches into attractive shapes. The objective is to provide sturdy, evenly shaped branches.

Arrange the blossoms evenly on the stem. With the garden clippers, trim off the protruding toothpicks. This picture illustrates the procedure with carrots.

Comments

Fresh ginger blossoms, called *miyoga* in Japanese, are available during the summer in Oriental food stores. They are often packaged pre-trimmed and only need minimal cleaning. Do not soak this flower in ice water. The gladiolus should be the last flower made for a bouquet.

Tricks

For the successful creation of this flower, the shape and size of the leek stem is important. The long leek used in this picture was purchased at an Oriental vegetable market.

Jicama Day Lily

Materials Needed

One firm, medium-size jicama or rutabaga, blueberries, toothpicks, and a bamboo skewer.

Tools Needed

A French knife, a paring knife, and garden clippers.

Place the vegetable on a cutting board with the root end up. With the French knife, make five vertical cuts. The cuts should be slightly slanted toward the center. The vegetable should have five sides, all slanting inward.

With the tip of the paring knife, make five vertical cuts into the vegetable to form petals. Be sure the petals remain attached. The petals should be slightly curved and pointed.

*C*arve out the center.

*I*nsert four to six toothpicks randomly into the center. Trim the toothpicks to approximately 1/8″ in length. Place one blueberry on each toothpick. Insert the bamboo skewer from the bottom.

Comments

This flower can be soaked in ice water without the blueberry center. Other root vegetables, such as white turnips, kohlrabi, or rutabaga, can be used to create this flower.

Tricks

The jicama, a South American vegetable, is rather firm. It is advisable to use a French knife to make the first cuts.

23

Leek Paint Brush Flower

Materials Needed

One thick, large-size leek, chives, and bamboo skewers.

Tools Needed

A paring knife.

Cut off the roots at the base. Be sure not to cut off too much of the base or the vegetable will fall apart. Cut off the top of the leek approximately 5" from the end of the root.

Hold the leek in your hand, or place it on a cutting board, with the root end down. Make several vertical cuts, but do not cut through the end of the root.

24

Open the flower.

Comments
Large-size leeks often have a solid center and are unsuitable. The flower is easy to make, and with a little practice, it can be made quickly. Instead of using bamboo skewers as stems, wicker sticks can be substituted.

Tricks
Store the flower in ice water for at least one hour for it to open fully.

Leek Flower

Cut the leek approximately 6" from the end of the root. Place the leek on your work table, with the root end pointing away from you. Carve long, slender petals with the tip of the paring knife.

Depending on the size of the vegetable, make approximately five to six petals. Make sure the knife tip is inserted all the way to the center. Remove the excess material between the petals.

26

Cut off the roots, but leave the base intact to hold the flower together.

Open the flower and cut the center leaves shorter. Insert the bamboo skewer.

Comments

Some large-size leeks have a woody, solid center. These are unsuitable for this application. Check the vegetables when purchasing them to avoid unpleasant surprises.

Tricks

The flower in the picture is shown with fresh chives to emphasize the attractive contrast between the white flower and its green surroundings. Store the flower in ice water for at least one hour for it to open fully.

Radicchio Flower

Remove any dry outer leaves.

Carefully break a few leaves away from the center core. The leaves may be tightly packed around the core. Be sure to keep the leaves intact. Do not destroy the center core. Open the core carefully and be sure the leaves remain attached to the core.

$\mathscr{A}$ttach the leaves that have been broken off to the core with toothpicks.

$\mathscr{S}$hape the flower as you work. It should look naturally unruly. Insert the bamboo skewer into the bottom.

$\mathscr{C}$omments

Radicchio is a purple lettuce with a firm head. Most radicchio is imported from Italy, but some is now grown in California. The height of the season is winter. This flower should not be soaked in ice water because it will eventually become water logged and limp. The color is partially water soluble.

$\mathscr{T}$ricks

The flower works best as a color highlight in large arrangements.

29

Radish Dahlia

Materials Needed

Several firm, large-size red radishes, tooth-picks, and a bamboo skewer.

Tools Needed

A paring knife and garden clippers.

Cut 10 to 12 thick, sturdy petals from the radishes. It will take three to four radishes to make one flower.

Select one radish as the center. Arrange petals on a cutting board in a circle to esti-mate the approximate size of the radish that you will need for the center. Take the center radish and, with the knife tip, make shallow, triangle-shaped incisions. Fold back the skin slightly to make the cuts resemble center petals.

30

With the garden clippers, cut the tooth-picks to approximately 1". Insert a long toothpick into the center petals, and make sure it is concealed. Secure the petals around the center. Insert the bamboo skewer into the base.

Comments

Do not soak this flower in ice water because it will change shape and loose color.

Tricks

Make the flower look as natural as possible by overlapping some of the petals. The beauty of the flower depends upon the freshness of the radishes. They should be crisp and a shiny red. Old radishes with crinkled skins look unattractive and will not fasten securely to the base.

Radish Cherry Blossom

Materials Needed

One firm, large-size red radish, one yellow squash, and a toothpick or bamboo skewer.

Tools Needed

A paring knife and a round and oval melon scoop.

Turn the radish with the root end down. With a paring knife tip cut 5 petals, starting from the top and working your way toward the bottom. Be sure the petals remain firmly attached to the base. Remove some of the center material with the knife tip.

Remove the additional material with the oval melon scoop. The center should be bowl shaped and as deep as possible.

Make tiny incisions around the leaf edges with the paring knife.

Cut out a yellow squash ball with the small melon scoop. Insert a toothpick or a bamboo skewer through the bottom of the radish flower and place the squash ball in the center.

Comments

This flower was photographed with daikon radish sprouts.

Tricks

With many blossoms arranged around a twig, the flower looks very attractive in arrangements. The choice of using toothpicks or bamboo skewers depends upon how the blossoms are used and the length of their stem. The flower can be preserved in ice water overnight. Some of the color will leak out, but this is not necessarily unattractive. Eventually the flower will become lighter in color and possibly even mottled.

Red Cabbage Decorative Leaves

Materials Needed
One firm, large-size red cabbage and tooth-picks or bamboo skewers.

Tools Needed
A paring knife.

Use the outer leaves to create the leaves. Place the cabbage leaves on a cutting board.

With the tip of the paring knife, cut various shapes from the fleshy part of the leaves near the ribs.

The ribs are paler than the rest of the leaves, and provide nice color contrasts.

Fasten the leaves to the flowers or to the slender leak stems with toothpicks.

Comments

These red cabbage leaves make nice filler pieces in arrangements.

Tricks

The leaves can be soaked in ice water, but the color may be somewhat water soluble. To partially remove cabbage stains from your hands, equipment, or linens, use lemon juice.

Red Cabbage Daisy

Materials Needed

One firm, medium or
large-size red cabbage,
one large-size
blueberry, and
a bamboo skewer.

Tools Needed

A paring knife and a
French knife
(optional).

Remove the wilted outer leaves. Place
the cabbage on a cutting board and cut off
a piece approximately one-quarter from
the top. If the cabbage is large, a French
knife should be used.

Place the cabbage on a cutting board with
the cavity facing upward. Remove the
smaller center leaves, keeping approximately four layers of leaves. The resulting
piece should be shaped like a saucer. With
the tip of the paring knife cut oval-shaped
petals. Be sure not to cut across the center.

36

Insert a bamboo skewer into the bottom of the flower. Slightly turn each layer of petals so that the petals are alternating. Place one large-size blueberry onto the skewer tip to create the flower's center.

Comments

Use young summer cabbage. The leaves do not have fleshy center ribs. This flower is rather fragile because it does not have a strong center core to hold it together. It might not remain upright very long on a bamboo skewer without additional support. Rest it against other flowers to provide support if needed.

Tricks

The flower can be kept in ice water without the blueberry, but the color is water soluble and will eventually leak out. This flower can be made in conjunction with Red Cabbage Decorative Leaves (see page 34), which use the lower, ribbed part of the cabbage.

37

Red Cabbage Large Flower

Materials Needed

One firm, medium or large-size red cabbage, blueberries, a bamboo skewer, and toothpicks.

Tools Needed

A paring knife, a French knife (optional), and garden clippers (optional).

Remove the wilted outer leaves. Place the cabbage on a cutting board and cut off a piece approximately one-quarter from the top. If the cabbage is large, a French knife should be used.

Place the cabbage upside down, with the cut portion resting on the cutting board. With the paring knife carve eight to ten petals around the cabbage. Make sure that they are held together by the cabbage trunk. Insert the knife as deep as possible because you will need to remove the center portion.

38

$\mathcal{P}$lace the cabbage on its side and carve out the center, but be careful not to cut the petals from the trunk. Remove the center completely.

$\mathcal{I}$nsert a bamboo skewer through the bottom. Insert eight to ten toothpicks into the center core. Shorten them if necessary so that they protrude approximately 1/8" to 1/4". Place one blueberry on each toothpick. Cluster the blueberries to create a substantial center.

$\mathcal{C}$omments

This flower is large and very attractive in large bouquets. The cabbage should be firm and fresh. The size of the cabbage obviously determines the size of the final flower. It is always heavy and needs a strong support.

$\mathcal{T}$ricks

This flower can be soaked without the blueberries in ice water. Red cabbage color is water soluble. A small amount of lemon juice added to the water will intensify the purplish color.

Red Cabbage or Carrot Bird of Paradise

Materials Needed

One medium-size red cabbage or one firm, large-size carrot, one green long-stemmed leek, and toothpicks.

Tools Needed

A paring knife, garden clippers, and a peeler (optional).

Use the outside cabbage leaves with fleshy ribs. Cut the cabbage leaves into slivers, or into triangular-shaped leaves, along the rib.

If a carrot is used, peel and cut it into sections approximately 4" long and then slice it into horizontal slices approximately 1/8" thick. Place the slices on a cutting board and trim the edges with tiny incisions to make V-shaped leaves.

$\mathcal{T}$rim the leek if necessary and shape the leaf ends into points.

$\mathcal{I}$nsert toothpicks into the cabbage blossoms (or carrot blossoms). Arrange the blossoms attractively on the leek. With the garden clippers, trim off any protruding toothpicks.

Comments

Do not soak the leek in water because this will cause it to soften. It is best not to mix cabbage flowers and carrot flowers on the same stem. The flower presented in this picture is only for illustration.

Tricks

The flower is most attractive when it is tall and in large arrangements. The shape and size of the leek stem are important for the successful creation of this flower. The long leek used in this picture was purchased at an Oriental vegetable market.

Red Onion Mum

Materials Needed

One medium-size red onion, chives (optional), and a bamboo skewer.

Tools Needed

A paring knife.

Place the onion on a cutting board with the root end down. Hold onto the top and make approximately 12 to 15 incisions toward the center approximately one-third of the way down. Spear inward, rather than slice; the incisions should be zigzag, creating triangular petals. Cut all the way to the middle of the onion.

Hold the onion at the top end and carefully peel away the root layers. Make sure the top layer remains together. This part will become the flower.

42

*C*arefully remove all of the layers and the onion skin. Keep the top layer together.

*I*f necessary, turn and twist the leaves carefully for additional dramatic effect. Spear them with a bamboo skewer.

Comments

Red onions are actually purplish and are very pretty, but they have a strong smell that may be irritating to some people. This flower looks stunning paired with a spray of chives. Handle with care because the flower is delicate. There is little support to hold it together. Do not store the finished flower in ice water.

Tricks

When making a bouquet of only mums, use different size onions to create a very natural effect.

Red Onion Flower

Materials Needed

One red onion, one red radish, one firm, large-size red raspberry, a bamboo skewer, and lemon leaf stems (optional).

Tools Needed

A paring knife.

Peel the onion but do not cut off the top; it should remain on with some skin. Cut the onion in half lengthwise, from the root end down toward the top. Place the onion on a cutting board with the cut side down and cut it into four wedges. The outer layers will become the flower petals.

Separate the layers and trim them into petal shapes. Do not cut away the tip; it is often darker than the rest and provides an attractive accent. About six to eight petals are needed for each flower.

44

Cut the radish in half. This will be the base. Spear the radish from the bottom with a bamboo skewer. The skewer tip should protrude approximately 1/2" above the flat side of the radish. Arrange the petals on the base, spearing each with the bamboo skewer. The darker ends of the petals should be on the outside.

Decorate the center with a large raspberry.

Comments

The flower is fragile and must be handled with care. Red onions have a strong smell that might be offensive to some people. Do not store finished flower in ice water.

Tricks

The raspberries used in the picture make a nice center, but when they are not in season, other red berries, such as cranberries, can be used. Also attractive are centers made with firm, red radishes, or with beet balls, scooped from whole, raw beets.

45

Red Pepper Poinsettia

Materials Needed

One firm red pepper, one small-size yellow squash, a bamboo skewer, and toothpicks.

Tools Needed

A paring knife, a very small (pea size) melon scoop, and garden clippers.

Place the pepper with the stem end down on the cutting board. Make vertical cuts along the ridges to get solid pieces that are as large as possible.

Place the pieces flat on the cutting board and cut them into pointed petals. You should get one to two petals from each piece. It is important to make the petals as large as possible. Do not press down hard; the petals might snap. Trim the petals as needed on the inside. Each flower should consist of seven to eight petals.

46

Cut one slice approximately 3/4" thick from the stem end of the squash. This piece must be solid because it will be the base. With the melon scoop cut pea-size disks from the remaining squash. The disks should be bright yellow. Approximately eight to ten disks are needed for each flower.

Fasten the petals, shiny skin side up, with toothpicks onto the flat base of the squash. The petals should meet in the center. Place one squash disk on each toothpick end. Clip off any protruding toothpicks. Insert the bamboo skewer from the bottom up.

Comments

The flowers should not be soaked in ice water because they can soften. Also, the little squash disks can easily loosen and fall off.

Tricks

The flowers look very real and pretty, especially when combined with gold ribbons and green leaves.

Red or Yellow Pepper Daisy

Materials Needed

One firm, large-size red or yellow pepper, one radish, one large-size blueberry, and a bamboo skewer.

Tools Needed

A paring knife, a serrated knife (optional), and garden clippers.

Place the pepper on the cutting board and cut off vertical slices as large as possible. Turn the slices around and remove any white membranes. Be careful not to break the slices, and leave them as thick as possible. One slice is needed for each flower.

Place a slice on the cutting board, with the shiny, outer side up. With the tip of the paring knife, cut eight to twelve petals. The pepper skin is rather tough and a sharp knife must be used. Be sure the petals remain attached in the center.

48

Cut the radish in half and insert the skewer. The skewer tip should protrude through the flat base. Place the pepper flower on top.

Finish the flower by placing the blueberry in the center, secured by the tip of the skewer. With the garden clippers, trim the skewer to a desired length.

Comments

The peppers must be large and fresh to get a large piece for the flower. Pepper skin is tough. If you do not have a very sharp knife, use a serrated knife to cut the pepper slices. Do not store the flowers in ice water. The components can be assembled quickly at the last moment.

Tricks

When blueberries are not in season, the center of the flower can be made with a piece of pepper stem cut horizontally. (In the picture, the red flower is made with a pepper stem center.) Peppers are also available in red, yellow, green, and purple. Very nice bouquets can be created by using different colored peppers.

Red or Yellow Pepper Large Flower

Materials Needed

One red or yellow pepper, one fresh pearl onion or red onion, one firm zucchini, and a bamboo skewer.

Tools Needed

A paring knife and garden clippers.

Remove the stem and place the pepper on the cutting board with the stem end down. Insert the tip of the paring knife and carve large petals, starting approximately 1" from the top. They must remain attached in the center.

Cut eight to ten pointed, slender petals. Carefully separate the upper part from the lower part. The part without seeds (not the stem end) will be the flower. It should be cup shaped.

50

*C*arefully remove all seeds and white membranes.

*C*ut one zucchini slice, approximately 3/4" thick which will become the base of the flower. Insert the bamboo skewer through the zucchini into the pepper. With the garden clippers, trim the skewer to a desired length. Add the center. It can be either an unpeeled pearl onion or the center core of a red onion. Spear the center on the protruding tip of the bamboo skewer.

Comments
Do not store these flowers in ice water.

Tricks
The choice of center depends on the artistic preference of the chef. Red onion hearts have a slightly purplish tip and provide a dramatic color accent. To make red onion centers, just peel the onion layers away until the center piece is just the right size. This flower can be made with different colored peppers; a bouquet of pepper flowers in different colors can be very attractive.

Turnip Carnation

Materials Needed

One firm white turnip, toothpicks, a rubber band, and a bamboo skewer (optional).

Tools Needed

An electric meat slicer, a paring knife, and garden clippers.

Peel the turnip.

Slice the turnip horizontally with an electric meat slicer set at #2. Make thin, round slices in different sizes. Approximately ten to twelve slices are needed for one flower.

52

Select the smallest slice and shape it into a center cone. Press the bottom together and ruffle the top.

Holding the center cone firmly, place the other slices around it. The tops should be curly and resemble a carnation, the bottom should be pressed to the center cone. Secure the slices with toothpicks while you work. Use as many slices as needed until the desired shape is achieved. Wind the rubber band around the base and with the garden clippers, trim off any protruding toothpicks.

Comments

This flower can be soaked in ice water. *Kanpyo* (thin Oriental dried gourd strips) can be used instead of rubber bands, but they will soften in water.

Tricks

A bamboo skewer can be used to prop up the flower, but it will not hold well because there is no solid center. The flower is best displayed close to a base for support.

53

White Turnip Daisy

One firm white turnip,
one yellow squash,
and chives.

Tools Needed
An electric meat
slicer, a vegetable
peeler (optional),
a paring knife, and
a small-size melon
scoop.

Peel the turnip. Slice it on the electric meat slicer set at #2. Make eight to ten thin, round slices. Stack the slices on top of each other. With the tip of the paring knife, cut out eight to ten pointed petals, but be sure the petals remain attached in the center.

Fan out the slices so the petals will alternate and make a nice display.

Cut a piece of squash 1/2" square and 1/4" thick. This will be the base. Spear it with a toothpick and be sure the yellow skin part is facing down. Place the turnip petals on top with a toothpick protruding approximately 1/4" through the center.

Using the melon scoop, carve out a round disk. Place the disk in the center of the flower.

Comments

Other root vegetables can be used, such as a red beet, a kohlrabi, a rutabaga or a large-size white daikon radish. This flower can be soaked in ice water, but the squash will eventually become waterlogged.

Tricks

If you make flowers with beets, be careful, because beet juice is very strong and will stain anything upon contact. To help remove beet stains from your hands and equipment, use lemon juice.

55

White Turnip Calla Lily

Materials Needed

One firm, large-size white turnip, fresh or canned yellow baby corn, toothpicks, and a bamboo skewer.

Tools Needed

An electric meat slicer, a vegetable peeler (optional), a paring knife, and garden clippers.

Peel the turnip (optional). Slice the turnip horizontally on the electric meat slicer set at #2 and make thin, round slices. One slice is needed for each flower. Start making the flowers by placing one corn in the center of a slice.

Wrap the lower part of the slice around the corn and secure it with a toothpick. Be sure the upper part is open and is folded back nicely.

Wrap the lower part of the slice around the corn and secure it with a toothpick. Be sure the upper part is open and is folded back nicely.

Spear the flower from the bottom with a bamboo skewer.

Comments

Do not soak this flower in water.

Tricks

If the turnip is firm and fresh, it does not have to be peeled. Some turnips have a purplish skin on part of the vegetable, and the purple rim on the slices can be very attractive. If the turnips are old, they should be peeled. Canned corn is easier to work with than fresh corn. Canned corn is salted and should be soaked to remove as much salt as possible. Salt that comes into contact with fresh vegetables causes them to become flabby and wilted.

Yellow Squash Sunflower

Materials Needed

One medium-size shiitake mushroom (grade A, approximately 2" to 3" across), one medium- to large-size yellow squash, a bamboo skewer, and toothpicks.

Tools Needed

A paring knife and garden clippers.

Cut off one 1" piece from the stem end of squash. This will be the flower base. Cut long, even slices from the vegetable. The slices should be as large as possible but not very thick.

Put slices on the cutting board with the skin sides up. Cut pointed petals, all approximately the same size. Eight to nine petals are needed for each flower.

*A*ssemble the petals with toothpicks on the flat side of the base of the squash. The petals should meet in the center.

*W*ith the garden clippers, cut off the toothpick ends to about 1/8". Insert the bamboo skewer through the bottom. Remove the stem from the mushroom and place the cap in the center, secured with the toothpick ends.

*C*omments

The squash should be firm to keep the petals in place. Fresh mushrooms work best, but dried mushrooms can also be used if soaked until soft and pliable. Do not store this flower in water.

*T*ricks

Yellow squash is often spotted. This is perfectly acceptable because nature is never perfect, and the brown spots blend well with the brown mushroom cap. The flower is stunning and can be made in different sizes.

Zucchini Tulip

Materials Needed

One firm, medium-size zucchini, a small-size leek paint brush flower used as the center (see page 24), and a bamboo skewer.

Tools Needed

A paring knife.

Hold the zucchini firmly, with the stem end toward you. Starting approximately 3″ from the stem, make five to six cuts with the paring knife toward you. Do not cut all the way through. These cuts will be the petals and must remain attached at the base.

Hold the flower base firmly and carefully twist the zucchini to separate the upper part from the carved flower.

Clean the center with the paring knife if necessary. It should be empty. Insert the bamboo skewer from the bottom.

Make a leek paint brush flower from a small-size leek (see page 24). Secure the center onto the bamboo skewer.

Comments

This flower is very attractive in arrangements.

Tricks

The zucchini should be very firm and dark green. The thinner the petals are, the more attractive the flower will be. The leek center should be made one day ahead of time and soaked in ice water. It will open up nicely. The finished flower should not be kept in ice water because the zucchini will get soggy.

Flower Centerpieces

Buffet Centerpiece

Created Using

Red pepper flowers
Cherry tomatoes
Yellow squash lilies
Red cabbage flowers with leek centers
Leek flowers
Eggplant flowers
Fresh rosemary
Blueberries
White turnip daisies
Lemon leaves
Fresh chives
Fresh sage
Rattan stems

Wicker Basket Supplied by

North Shore Floral, 300 Woodbury Road, Woodbury, NY 11797,
(516) 357-1600
Photographed in the Park Avenue Lobby

Wedding Centerpiece

Created Using

Calla lilies
Collard greens
Small green chili peppers
Fresh dill
Leeks
Lemon grass
Wicker stems
Rattan stems

Vase Supplied by

The Waldorf-Astoria
Photographed in the Conrad Suite Reception Room

Father's Day/Oriental Motif

Created Using

Kale leaves tied with kanpyo (Oriental dried gourd ribbons)
Leek mums
Carrot gladiolus on leek stems
Carrot paint brush flowers
Red cabbage leaves
Pine cones
Wicker stems
Kale stems
Eggplant daisies with raspberry centers
Fresh rosemary
Leek stems
Oriental leek stems

Bowl Supplied by

Inja Nam
Photographed in the Hilton Room

Valentine's Day

Created Using

Beet roses on lemon leaf stems
Whole artichokes
Wicker stems with raspberries
Beet pearls on stems

Dishes Supplied by

The Waldorf-Astoria Florist
Photographed in the Cole Porter Suite

Outdoor Party with Fruits and Berries

Created Using

Acorn squash halves filled with raspberries, blueberries, and blackberries
Yellow pepper daisies
Mirliton squash (also called chayote) squash
Kiwi
Figs
Nectarines
Strawberries
Tiger avocado (a special avocado available at Oriental markets)
Chinese squash
Fresh thyme leaves
Lemon leaves

Platter Supplied by

Inja Nam, made by Artcraft Studios
Props supplied by The Waldorf-Astoria Flower Shop
Photographed in The Waldorf-Astoria Flower Shop

Silver Anniversary

Top Tier

Purple onion flowers with
blackberry centers
Carrot leaves

Middle Tier

Turnip tulips
Zucchini tulips with
raspberry centers
Carrot tulips with leeks
Beet stems

Bottom Tier

Radish flowers with
yellow squash centers
Brussels sprouts
Asparagus
Shiitaki mushrooms
Pearl onions
Fresh asparagus

Handle

Yellow turnip flowers
with green pea center
Lemon leaves

Three Tier Silver Compotier Stand Supplied by

The Waldorf-Astoria
Photographed in the Silver Corridor in front of an elevator door

Birthday Party

Created Using

Bird of paradise flowers
Garlic sprouts (nira in Japanese)
Long Chinese green squash
Mung bean sprouts
Rattan stems

Vase Supplied by

Inja Nam
Photographed on the Ballroom checkroom counter

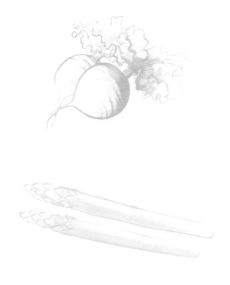

Thanksgiving

Squash Vase

Eggplant flowers with pearl onion center
Rosemary
Carrot tulips with raspberry centers
Pumpkin flowers with blueberries
Leek flowers
Red cabbage flowers
Red cabbage leaves
Lemon grass with cranberries
Lemon leaves
Parsley

Large Ceramic Vase

Dried wheat and straw
Dried Chinese lanterns
Dried poppies pods
Fresh bay leaves
Rattan stems
Fresh ginger flowers
Red pepper flowers with blueberry centers

Coconut Basket

Gorbo (Burdock)
Red delicious apples
Granny Smith apples
Chinese long squash
Acorn squash
Mirliton squash
Gourd
Jicama
Brussels sprouts

Other Decorations

Indian corns
Anjou pears
Lemon leaves
Basket filled with cranberries
Calabasa squash
Red grapes
Rattan stems

Vase and Basket Supplied by

Inja Nam
Photographed on the Starlight Roof Balcony overlooking Park Avenue

Independence Day

Blue Tray

Pearl onions
Raspberries
Blueberries

Firework Vase

Red pepper flowers with blueberry centers
Leek flowers
Pearl onions on wicker stems
Red cabbage leaves on rattan stems
Chives
Radish sprouts
Parsley

Vase and Tray Supplied by

Inja Nam
Photographed on the Ballroom East Foyer Balcony

Big Lobster Display

Silver Platter

Endive leaves filled with lobster salad, garnished with
lobster claws and tarragon leaves
Red onion flowers (water lilies) with raspberry centers
Cucumber slices

Background

White onion flowers
Pearl onions
Gourd
Lobster (approximately 28 lbs.)
Seaweed
Pearl onion flowers
Radish sprouts

Silver Platter Supplied by

The Waldorf-Astoria
Photographed in the Main Kitchen at the Sauce Cook Station

Easter Bonnets

White Hat

Acorn squash carnation
Fresh bay leaves
Fresh litchi nuts

Yellow Hat

Calla lilies
Chives

Pink Hat

Radish flowers with yellow pepper centers
Fresh rosemary

Hats Supplied by

Inja Nam
Photographed in the Vertes Suite

Spring Bouquet

Created Using

Bermuda onion flowers
Yellow squash flowers with onion centers
Lemon leaves
Dill
Lemon leaf stems

Bowl Supplied by

Inja Nam
Photographed in the Hoover Suite Foyer

Tray with Assorted Canapés and Flowers

Canapés

Demitasse spoons filled with half quail eggs, salmon caviar, cucumbers
Sturgeon strips wrapped in seaweed (Nori), radish sprout garnish
Lobster-tail slices decorated with lobster claws on rye bread, fresh thyme garnish
Smoked salmon and cream cheese on pumpernickel, radish sprout stems
Zucchini tulips with crab salad, papaya garnish
Smoked chicken, red peppercorn, and zucchini leaf garnish on whole wheat bread
Foie gras (goose liver) paté with truffle garnish

Large Vase

Yellow squash flowers with red cabbage centers
Chives
Fresh rosemary

Bouquet

Yellow squash flowers
Leeks
Chives

Vase Supplied by

Inja Nam
Photographed in the Ballroom East Foyer

Summer Bouquet

Created Using

Leeks
Brussels sprouts
Chervil
Rosemary
Fresh thyme
Fresh bay leaves
Mirliton squash
Gorbo (Burdock)
Shiitaki mushrooms
Asparagus
Ginger buds
Radicchio
Carrot flowers
Rattan sticks
Red cabbage flowers with pearl onion centers
Lemon grass
Yucca root
Ornament twigs
Yellow pepper flowers with raspberry centers

Bowl Supplied by

Inja Nam
Photographed in the Ballroom coat check corridor

New Year's Day

From Left to Right, First Row

Radish flowers
Onion flowers with pearl onion centers
White cabbage flower with blackberry center
Fresh litchi nuts
Chinese long beans
Pearl onions
Eggplant flowers with leek centers
Leek flowers
Leeks

Second Row

Eggplant daisies with pearl onions
Red cabbage flower with blueberries
Yellow squash flowers with blackberry centers
White turnip flowers with garlic chive blossoms

Third Row

Pearl onions
Jicama flower with blackberry
Acorn squash sunflower with shiitaki mushroom center

Fourth Row

Red cabbage flowers with garlic chive blossoms
Red beets on rattan stems with cranberry centers
Lemon leaves
Parsley
Rattan stems

Silver Punch Bowl Supplied by

The Waldorf Astoria
Photographed in the Silver Corridor

Oriental Arrangement

Created Using

Fresh lotus roots
Chinese bean sprouts
Bok choy flower and leaves
Acorn squash sunflowers with shiitaki mushroom centers
Chives with blossoms
Garlic chives with buds
Enoki (dake) mushrooms
Yellow pepper flowers with hot pepper centers
Yellow squash flowers with hot pepper centers
Branches
Lemon leaves

Chinese Bowl Supplied by

Inja Nam
Photographed in the Park Avenue Lobby

Christmas Centerpiece

Created Using

Red pepper flowers with yellow squash center
Red pepper rounds on wicker stems
Fresh thyme
Fresh rosemary
Pine cone wreath
Gold ribbon
Gold ornaments

Table Setting with Gold China Supplied by

The Waldorf-Astoria
Photographed in the Conrad Suite

Golden Anniversary

Created Using

Yellow pepper flowers with red onion (Bermuda onion) centers
Lemon leaves

Chung Dynasty Bowl Supplied by

Inja Nam
Photographed in the Basildon Room

Engagement Bouquet

Left Vase

Radish flowers with yellow pepper centers
Red onion (Bermuda onion) centers
Chives
Lemon leaves
Twigs

Right Vase

Carrot flowers
Yellow squash flowers with blueberry and turnip centers
Chives
Twigs
Rattan stems

Vases Supplied by

Inja Nam
Photographed in the Park Avenue Suite

Crudité Assortment on a Silver Tray

Back Row

Celery
Green pepper
Carrots
Scallion
Zucchini
White turnip

Front Row

Asparagus
Radishes
Endives
Red pepper
Snowpeas
Mushrooms
Cucumbers

Center Row

Enoki mushrooms
Broccoli
Cauliflower
Yellow pepper
Fennel
Cherry tomatoes with carrot balls

Dips from Left to Right

Dill in sour cream
Curry mayonnaise
Russian dressing

Silver Equipment Supplied by

The Waldorf-Astoria
Photographed in the Silver Room

Small Centerpiece

Created Using

Artichokes
Asparagus
Red cabbage leaves
Leek flowers
Chives
Endive hearts
Red beet flowers
Rattan sticks
Bamboo skewers

Jae - Seuk Kim Vase Supplied by

Inja Nam
Photographed in the Astor Gallery

Mother's Day

Created Using

Red pepper flowers with pearl onion centers
Turnip flowers with baby corn
Turnip daisies with yellow squash centers
Chives
Rosemary
Lemon leaves
Dried twigs
Dried poppies
Bamboo poles

Vase Supplied by

Inja Nam
Photographed in the Starlight Roof